P9-DNF-861

Noble Gases

Jens Thomas

MARSHALL CAVENDISH
NEW YORK

Benchmark Books
Marshall Cavendish
99 White Plains Road
Tarrytown, New York 10591

www.marshallcavendish.us

Library of Congress Cataloging-in-Publication Data

Thomas, Jens.
Noble Gases / by Jens Thomas.
p. cm. — (The elements)
Includes index.
1. What are the noble gases? — 2. Where are the noble gases found? —
3. How were the noble gases discovered? — 4. How do we isolate the
noble gases? — 5. Special characteristics — 6. How do the noble gases
react? — 7. Liquid helium — 8. Lighting up our lives — 9. Lasers —
10. Unreactive atmospheres — 11. Helium underwater — 12. Radon in
medicine — 13. Chemical reactions.
Summary: Explores the history of the noble gases and explains their chemistry, their uses, and
their importance in our lives.
ISBN 0-7614-1462-2
1. Gases, Rare—Juvenile literature. [1. Gases, Rare.] I. Title. II.
Elements (Benchmark Books)
QD162.T48 2003
546'.75—dc21 2002000210

Printed in China

Picture credits
Front cover: James A. Sugar at Corbis
Back cover: Science Photo Library/Rosenfeld Images Ltd.
Photos courtesy of Air Products and Chemicals, Inc.: 17, 24
Professor Neil Bartlett: 14
Corbis: James A. Sugar 4; Roger Ressmeyer 25
CryoService Ltd: 12
DaimlerChrysler: 20
DESY: 16, 18
Image Bank: Mitchell Funk *i*, 19
Jozef Stefan Institute: Prof. Dr. Boris Zemva 30
Lawrence Berkeley National Laboratory: 13, 15
NASA: 6
NewsCast: Boots Company plc 23 (*bottom*)
NOAA: OAR/NURP; University of North Carolina, Wilmington 26
PA Photos: Michael Crabtree 21
PhotoDisc: InterNetwork Media 7
Science & Society Picture Library: Science Museum 8 (*top*), 8 (*bottom*)
Science Photo Library: Charles D. Winters 5; Clinique Ste Catherine/CNRI 27; J.C. Revy 11;
Orville Andrews 22; Rosenfeld Images Ltd. 10
Siemens press picture: *iii*, 23 (*top*)

Series created by Brown Partworks Ltd.
Designed by Sarah Williams

Contents

What are the noble gases?

The noble gases are a group of six chemical elements—helium, neon, argon, krypton, xenon, and radon—in the last column of the periodic table. When all the noble gases were identified, around one hundred years ago, chemists thought that they were very unusual. The noble gases could not be made to react with other chemicals, so they did not have many uses. Now that chemists know more about the noble gases, they can be used in many ways. Some of the gases light up fluorescent lamps, while helium is used to inflate balloons and blimps. Radon is even used to treat cancer.

Noble gas atoms

Everything around you consists of tiny particles called atoms. Inside each atom are even smaller particles called protons, neutrons, and electrons. The protons and neutrons cluster together to form the nucleus at the center of each atom. The electrons encircle the nucleus like planets

Helium is not as dense as air and does not react with other chemicals. These properties make the gas ideal for inflating balloons and blimps.

DID YOU KNOW?

ROYAL GASES

The noble gases have been called a number of names since they were first discovered. At first, people called the noble gases "rare gases" since they were so hard to obtain. They were also often called "inert gases" because of their reluctance to react with other chemicals. Today, however, some noble gases have been made to react with other substances so the term *inert* no longer applies. "Noble gases" is a much better description of these gases. The word *noble* explains that, like the kings, queens, and lords of nobility, the noble gases do not interact with other elements unless they are forced to do so.

orbiting the Sun. Groups of electrons revolve around the nucleus in layers that are like the rings of an onion. Each "layer" is called a shell, and it can only hold a certain number of electrons.

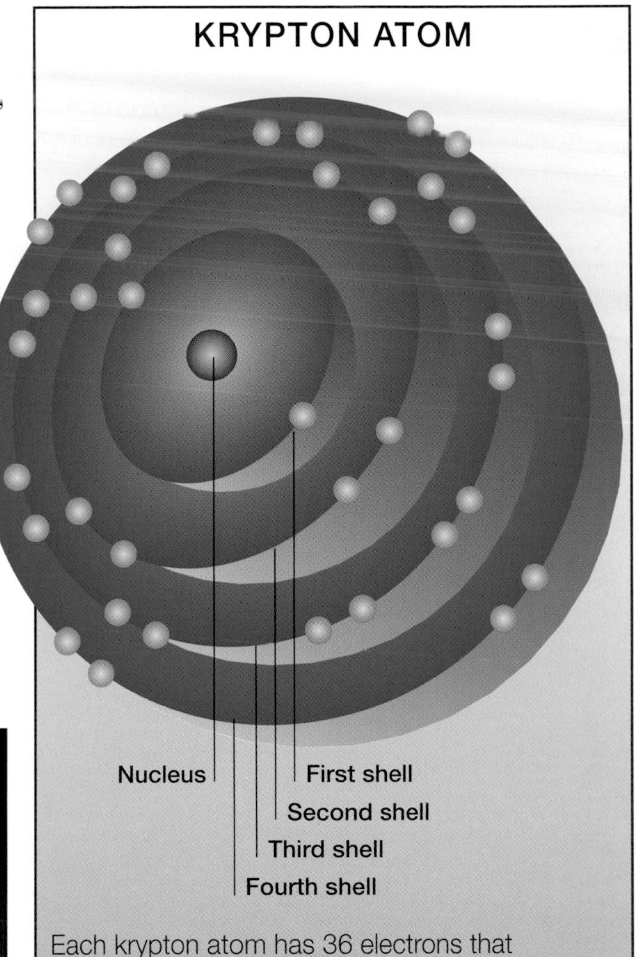

Each krypton atom has 36 electrons that revolve around the nucleus in 4 shells. There are 2 electrons in the first shell, 8 in the second, 18 in the third, and 8 in the outer shell. Each noble gas atom contains a different number of electrons and electron shells. Radon, for example, has 86 electrons orbiting the nucleus in 6 electron shells.

The chemistry of the noble gases

The noble gases are different from all other elements because they have enough electrons to fill their outer electron shell. All the other elements either have too few or too many electrons to fill up the outer shell. As a result, most elements take part in chemical reactions to gain or lose electrons. The full outer shells of the noble gases make them very unreactive. It also means that they do not behave like other elements.

An argon light glows pale purple as electricity flows through the gas-filled tube. Other noble gases glow different colors when they are put in the tube.

Earth's atmosphere contains trace amounts of argon, neon, and helium, and even smaller quantities of krypton, radon, and xenon.

Where are the noble gases found?

Take a deep breath and hold it in. You now have some noble gases inside your body. Most noble gases in the atmosphere have been around since Earth was created some 4.6 billion years ago. However, helium, argon, and radon are still being made. These noble gases form when unstable, or radioactive, chemical elements in underground rocks (such as uranium and thorium) break down. Some of these gases seep up through cracks in the rocks and escape into the atmosphere.

After hydrogen, helium is the second most common element in the Universe. It is especially common in stars and makes up about a quarter of the mass of our own star—the Sun. Indeed, helium takes its name from the Greek word *helios,* meaning "sun." Helium is a rare gas in the atmosphere. It is so light that Earth's gravity cannot hold onto it, and the gas gradually drifts off into space. Other noble gases can be found all over the Universe, but they are not as common as helium.

Under the ground

On Earth, the noble gases can be found trapped underground. Helium deposits are often found in reservoirs of natural gas and oil. The gas becomes trapped under an airtight covering of rock. The amount of helium found deep under the surface of Earth is around 3,000 times greater than the amount found in the atmosphere. Other noble gases are often found alongside coal and oil deposits. Tiny amounts of noble gases were trapped inside plants and animals when they died hundreds of millions of years ago. Over time, the gradual decay and compression of these remains have changed them into coal and oil.

DID YOU KNOW?

RADIOACTIVITY

Imagine leaving a sample of an element in a box somewhere safe. If you opened the box a few weeks later, you would be amazed if you found a completely different element in the box. This would happen if you put a sample of some radioactive elements in the box.

The nuclei of all radioactive atoms are unstable. These atoms try to become stable by spitting out particles in a process called "radioactive decay." There are three types of radioactive decay: alpha radiation, beta radiation, and gamma radiation. In alpha radiation, the radioactive atom spits out an alpha particle. This particle is the same as the nucleus of a helium atom. Once the alpha particle is released, it picks up electrons in the atmosphere and becomes an atom of helium gas. If a uranium atom loses an alpha particle, for example, the atom will become a thorium atom. The radioactive decay continues and the original uranium atom eventually becomes a lead atom.

Noble gases trapped underground are sometimes released into the atmosphere by erupting volcanoes and hot springs.

How were the noble gases discovered?

The first person to show that a noble gas existed was British scientist Henry Cavendish (1731–1810). In 1785, Cavendish passed electricity through a sample of air from which he had removed carbon dioxide gas. At the time, chemists thought that air only contained carbon dioxide, nitrogen, and oxygen. Cavendish thought that the electricity should have joined the nitrogen and oxygen gases to create a liquid called nitrous acid (HNO_3).

Sir William Ramsay identified the existence of all the noble gases on Earth except for radon. He was awarded the 1904 Nobel Prize for chemistry in recognition of his achievement.

No matter what Cavendish did, however, a small bubble of an unreactive gas remained. His discovery could not be explained for one hundred years. Then, in 1894, British scientists Sir William Ramsay

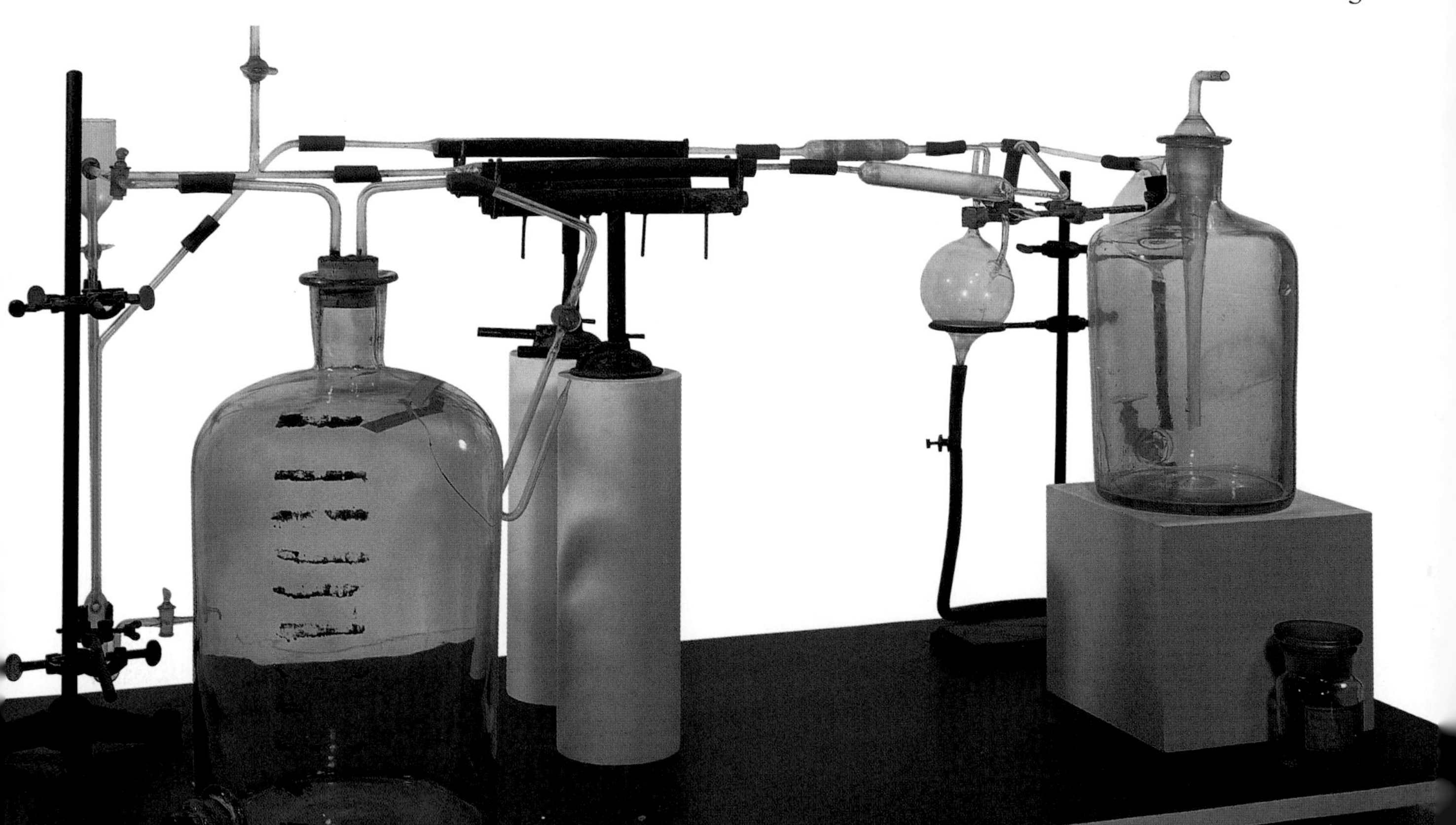

This is a replica of the apparatus used by British chemist Sir William Ramsay during his discovery of argon in 1894.

(1852–1919) and Baron Rayleigh (John William Strutt; 1842–1919) provided the answer. Both scientists found that when they removed nitrogen from the air, they were left with a bubble of gas. Ramsay realized that the gas must be a new chemical element. Since the new gas was so unreactive, Ramsay called it argon from the Greek word *argos,* meaning "lazy."

Ramsay immediately tried to find more noble gases. Four years later, together with British chemist Morris William Travers (1872–1961), Ramsay isolated three more noble gases—krypton, neon, and xenon.

First and last

Helium is the only element discovered in space before it was found on Earth. On August 18, 1868, British scientist Sir Joseph Lockyer (1836–1920) was studying a solar eclipse using a spectroscope. A spectroscope identifies elements by the spectrum of lines they produce when heated. Lockyer noticed a new line in the spectrum of the Sun. He decided that this must be due to a new element and named it helium. In 1895, Ramsay identified helium on Earth when he noticed it as a gas given off during the radioactive decay of uranium.

The last of the noble gases to be discovered was radon. This radioactive element was discovered in 1900 by German scientist Friedrich Ernst Dorn (1848–1916).

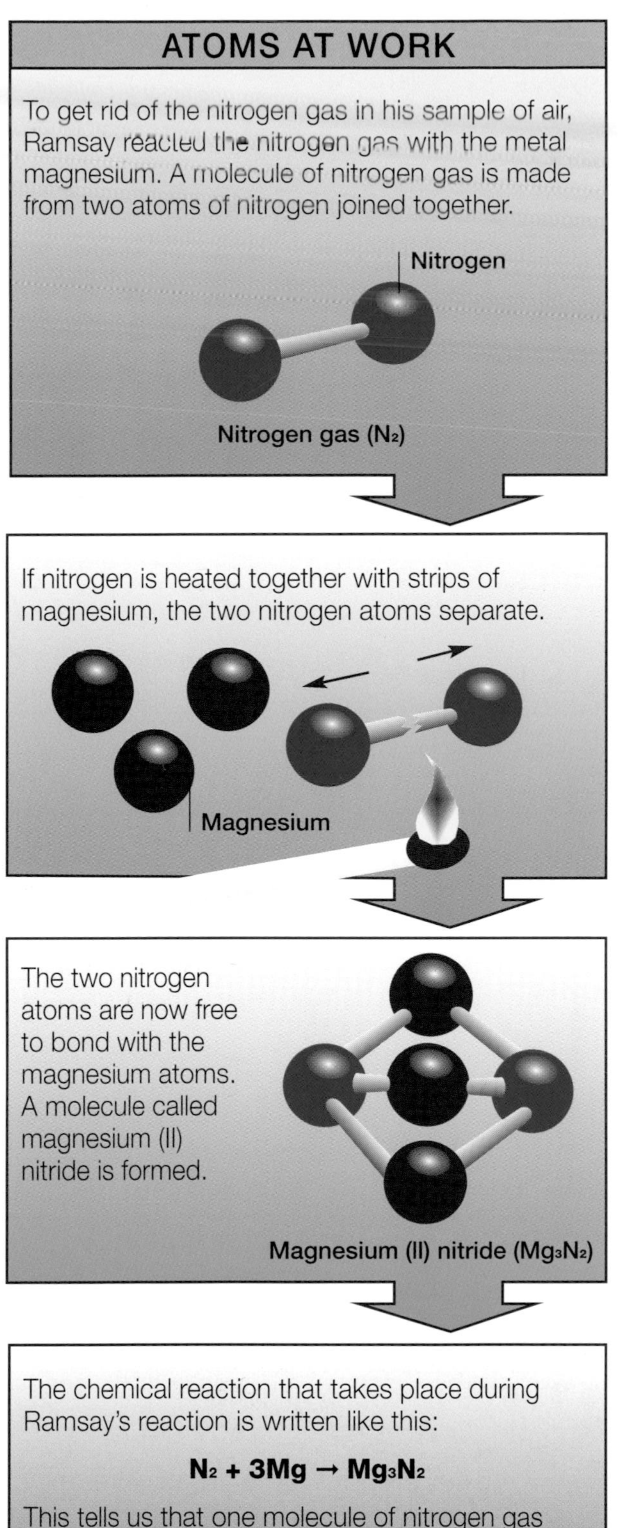

How do we isolate the noble gases?

Distillation towers at a chemical plant in the United States are used to separate the gases that make up air.

The noble gases are very difficult to separate from air. Air also contains oxygen and nitrogen, which are much more important in industry than the noble gases. For this reason, factories that make noble gases usually make oxygen and nitrogen as their main products.

Air is treated chemically several times before all the different gases are collected. First, the air is compressed to about six times its normal pressure. The air is then cooled so that any water, carbon dioxide, and other impurities in it freeze into ice crystals, which can be removed easily. The purified air is then cooled again until it is extremely cold. This turns all the gases in the air into liquids.

This liquid air is then pumped into a distillation tower. Here, some of the liquid nitrogen is removed. This leaves a liquid containing all the gases from the air, less nitrogen, and more oxygen. The liquid air is fed into a second distillation tower at a lower pressure. In this tower, gases in the liquid air boil off from the mixture and cool back into liquids at different points up the distillation tower. Xenon and krypton collect at the bottom of

DID YOU KNOW?

DISTILLATION

If you heat a liquid, the molecules in the liquid will eventually escape, or evaporate, to form a gas. This occurs at the liquid's "boiling point." Heat a liquid above the boiling point and it evaporates into a gas. Cool the gas below the boiling point and it turns back into a liquid.

Distillation is a process that separates a mixture of liquids with different boiling points. The liquid mixture is heated in a tower that is hotter at the bottom than it is at the top. Gases boil off from the liquid mixture and rise up the distillation tower. Each gas turns back into a liquid when the temperature in the tower drops below its boiling point. In this way, many different liquids can be separated from a single mixture.

the tower. Neon and argon collect farther up the tower. Each gas is drained and distilled again in different towers to purify them.

Helium is not usually collected by the distillation of air. Most helium comes from underground reservoirs of natural gas. Helium-rich deposits of natural gas are found in certain parts of the United States, most notably in the Rocky Mountains.

Radon gas is usually collected as it is released during the radioactive decay of minerals such as uranite (shown below) or pitchblende.

Special characteristics

At normal room temperature, all the noble gases are colorless and have no smell. Radon is the heaviest noble gas and helium is the lightest noble gas. After hydrogen, helium is the second lightest element of all. If the noble gases are cooled or squeezed under pressure, they become liquids. The temperature at which a noble gas becomes a gas is called its boiling point. The boiling points of the noble gases are very low. Radon has the highest boiling point of all the noble gases. It turns into a gas when the temperature has reached a chilly –79.74° F (–62.08° C). Helium has the lowest boiling point of any known substance. It turns into a gas at –452.09° F (–268.94° C). This is only 7.58° F (4.21° C) higher than absolute zero, which is the lowest temperature possible (–459.67° F, or –273.15° C).

DID YOU KNOW?

THE PERIODIC TABLE

In the late nineteenth century, Russian scientist Dmitry Mendeleyev (1834–1907) arranged all the known chemical elements into a chart called the periodic table. Mendeleyev grouped elements into rows and columns based on how they reacted and how many protons they contained. He left gaps for elements yet to be discovered.

The position of an element on the periodic table tells scientists how the element will behave. When the noble gases were identified, scientists thought that they would fit into gaps left in the table. But the noble gases were so different that a whole new column had to be created. This changed the periodic table forever.

To keep argon as a liquid, the element must be stored in high-pressure cylinders at temperatures of around –297° F (–183° C).

U.S. chemist Luis Walter Alvarez (1911–1988) isolated the only isotope of helium in the late 1930s.

If liquid noble gases are cooled or compressed, they become solids. Helium is very difficult to solidify. It becomes a solid at –457.87° F (–272.15° C) under extreme pressure. As liquids or solids, most noble gases are colorless. Radon is an exception. If liquid radon is placed in glass, it glows blue, blue-green, or lilac. Solid radon is even more exciting—it changes color over time. First it glows blue, then yellow, and finally orange-red.

Noble gas isotopes

All the noble gases exist in different forms called isotopes. Isotopes are atoms that have the same numbers of protons and electrons, but different numbers of neutrons. Isotopes of the same element react in the same way but have different weights. Many are also radioactive. Helium and argon have just one isotope. Neon has six isotopes, three of which are radioactive. All radon isotopes are radioactive.

NOBLE GAS FACTS

	Boiling point
Helium	–452.09° F (–268.94° C)
Neon	–410.89° F (–246.05° C)
Argon	–302.58° F (–185.88° C)
Krypton	–242.23° F (–152.35° C)
Xenon	–160.78° F (–107.10° C)
Radon	–79.74° F (–62.08° C)

How do the noble gases react?

For many years, most chemists thought that the noble gases would not react with any other substance. This behavior challenged the scientific world. So toward the middle of the twentieth century, many scientists tried to make compounds containing a noble gas.

Atoms form bonds by gaining, losing, or sharing electrons with other atoms. In most cases, only the electrons in the outer electron shell are involved in the reaction. Electrons in the outer shells of heavy atoms are far away from the nucleus, so the attractive force between the nucleus and these electrons is weak. This makes heavier elements more likely to bond with other elements. As a result, the work of most scientists centered around heavy noble gases such as xenon and radon.

After years of painstaking research, the scientists' dreams became a reality. Canadian chemist Neil Bartlett (1932–) made the breakthrough at the University of Columbia in Vancouver, Canada, in 1962. Bartlett was experimenting with a chemical compound called platinum hexafluoride, or platinum (VI) fluoride (PtF_6). During his research, Bartlett noticed that this substance changed color in air. This was a sign that the chemical was so reactive that it could take an electron from the oxygen in the air, letting the platinum compound and oxygen react with one another. Bartlett remembered that it was just as difficult to pull an electron from xenon as it was from oxygen. Bartlett tried mixing the bright red PtF_6 gas with colorless xenon gas. An orange-yellow solid immediately formed, showing that a reaction had taken place. A complex compound called xenon platinofluoride ($XePtF_6$) had formed.

Canadian scientist Neil Bartlett at his laboratory at the University of Columbia in Vancouver, Canada.

Bartlett's success led to the discovery of many other noble gas compounds, including various fluorine and oxygen compounds of xenon and a compound containing krypton and fluorine. However, no one has yet managed to make helium, argon, or neon react.

Clathrate compounds

Some noble gases form special compounds called "clathrates." These are made when a chemical called quinol or water are frozen in a container containing argon, krypton, or xenon at high pressure. Under pressure, the gas molecules are "squeezed" into the gaps between the quinol or water molecules as they freeze. Clathrates are a useful way of storing noble gases. However, they are not proper compounds, because no reaction actually takes place and the noble gas atoms remain unchanged.

Platinum hexafluoride (PtF_6) is a bright red gas (at left) that reacts with xenon to produce an orange-yellow compound called xenon platinofluoride ($XePtF_6$; at right).

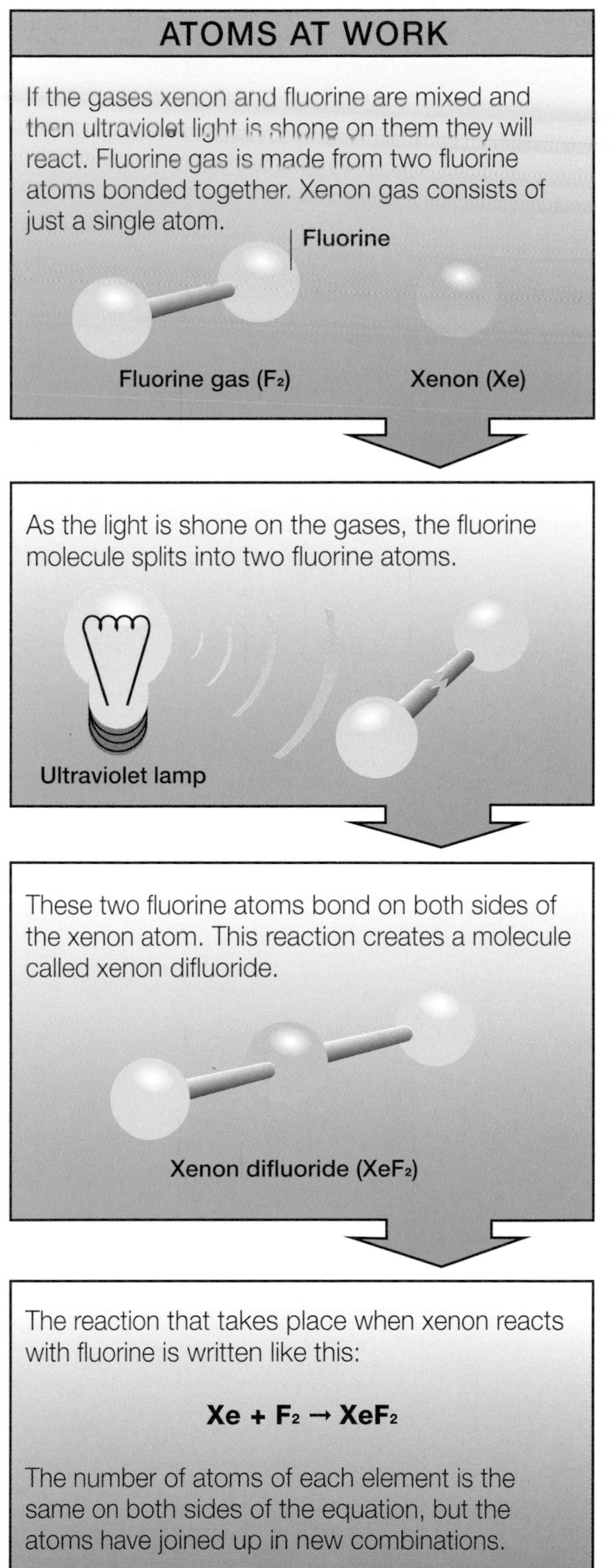

Liquid helium

Imagine that you stirred a hot drink and the liquid continued to swirl forever. Or what if you stepped into a bathtub and the water crawled up your legs and covered your entire body? This is the bizarre world that exists for liquid helium.

To become a liquid, helium must be cooled to –452.09° F (–268.94° C). The liquid at this temperature is called helium I. It is colorless and so transparent (see-through) that it is impossible to see the surface of the liquid with the naked eye.

Helium I exists just a few degrees above "absolute zero"—the coldest possible temperature. Most scientists measure temperature in units called Kelvins (K), after British scientist Lord Kelvin (1824–1907). Absolute zero is 0 K (–459.67° F, or –273.15° C).

Superfluid helium

Helium I always bubbles. If it is cooled to just 2.2 K above absolute zero, however, it becomes completely still and expands. It has now become helium II—a liquid that behaves in ways that are so different from

Liquid helium cools a ceramic superconductor, which floats on a magnetic field. Superconductors offer no resistance to an electric current, but they only work at extremely low temperatures.

other fluids that it is often called a "superfluid." If you lower an upright, empty test-tube half way into a bath of helium II, the liquid helium will "pour" into it until the level of liquid inside the tube matches the level in the bath. Take the half-filled tube out of the bath, and the helium in the tube will crawl up the walls of the tube and drip back down into the bath until the tube is empty.

DID YOU KNOW?

HOW LIQUID HELIUM IS MADE

To turn helium into a liquid, the gas is first cooled with liquid nitrogen or oxygen. Once it has been cooled to the temperature of these liquids, the helium must cool itself. This is achieved by expanding the gas, which pulls the atoms apart, slows them down, and cools the gas further. A cycle is then set up with expanded helium used to cool unexpanded helium, which is then itself expanded to become even colder. After many cycles, the helium becomes a liquid. Danish scientist Heike Kamerlingh Onnes (1853–1926) was the first person to liquefy helium in 1911.

Liquid helium cools the magnets in a magnetic resonance imaging (MRI) device. An MRI device uses magnetism to take pictures of things inside the body.

Powerful superconductors, cooled by liquid helium, hold beams of subatomic particles on course at the Deutsches Elektronen-Synchrotron (DESY) particle accelerator in Hamburg, Germany.

Another thing that makes helium II a "superfluid" is its viscosity. Viscosity is a measure of how "gooey" a liquid is. Oil has a higher viscosity than water, because oil is harder to pour than water and does not flow as easily. Helium II has zero viscosity, which means that it flows very easily. This is why liquid helium will keep on swirling once it has been stirred.

Understanding science

Atoms, and the particles they are made from, are too small to be seen, even with the most powerful microscopes. However, scientists do know that they do not behave like the objects we can see around us. Some particles can pass through solid barriers, while others appear to be in two places at the same time. Liquid helium is special because it is ruled by laws that let its tiny particles act in odd ways. Scientists can learn more about how these tiny particles behave by studying liquid helium.

Lighting up our lives

All the noble gases are colorless, so they are invisible to the naked eye. However, if you have ever seen a colorful glowing tube of light in a store's front window or outside a restaurant, then you have seen the glow that some noble gases can emit. These glowing tubes are called neon lights. They have been lighting up towns and cities throughout the world since their invention in the 1920s. Neon lights fell out of favor during the 1960s and 1970s, but they are increasingly popular today. In fact, neon lights are sometimes used in works of art.

A street in Las Vegas in the United States is illuminated by hundreds of glowing neon lights.

Making neon light

A neon light consists of a glass tube filled with a noble gas at a very low pressure. As the name suggests, neon is the most common gas used to make neon lights. Neon lights containing the other noble gases (aside from radon) are also common. Metal plates called electrodes are fitted to each end of the gas-filled tube. When the neon light is plugged into an electricity supply, an electrical current flows through the tube between the electrodes. In some lights, the voltage in the neon tube may be

The headlights of many automobiles are now filled with xenon, which makes the light very bright.

DID YOU KNOW?

BRIGHT LIGHTS

Some noble gases emit extremely bright light. Krypton is used when a very intense flash of light is needed, such as in the bulbs used for high-speed photography. High-intensity xenon bulbs are widely used as the source of light for lighthouses and rock concerts.

as high as 15,000 V. A battery in a clock or personal stereo supplies 1.5 V, so 15,000 V is the same as the power produced by 10,000 batteries. This huge surge of electricity excites electrons in the noble gas atoms. Electrons jump from electron shells close to the nucleus to electron shells farther away from the nucleus. Eventually, they go back to their original position. As they return, they release the extra energy as light. All the light you see is a result of electrons jumping between electron shells. Similarly, a flash of lightning in the sky occurs as electricity shoots through gas molecules in the air. The electrons in the air jump between different shells and release the energy as light. So the next time you look at a neon light, you are really watching miniature lightning.

Colored gases

Lightning is a bluish-white color because the air is a mixture of many different gases. The neon gas that gives neon lights their name glows fiery-red. But different gases glow different colors when they are excited by electricity. For example, helium atoms glow gold, krypton glows dull green, argon glows pale purple, and xenon glows bluish-gray. Many more colors can be made by mixing these gases together or by mixing them with the vapor of mercury and argon, which gives a bluish light.

Krypton-filled bulbs provide the intense burst of bright light needed for nighttime photography.

DID YOU KNOW?

FLUORESCENT LIGHTS

The white tube lights you may have at school are not actually neon lights—they are called fluorescent lights. The two lights differ in the way that the light is produced. In a fluorescent light, the "gas" is a vapor of mercury. When electricity passes through the vapor, the mercury atoms generate a type of light called ultraviolet light. People cannot see ultraviolet light unless the inside of the tube is coated with phosphorus. This changes ultraviolet light into visible light. To add to the confusion, many neon lights also contain mercury and have phosphorus coatings. Different colored lights have been made available by adding mercury to neon lights.

Mirrors reflect the intense green beam of a powerful argon ion laser.

Lasers

A laser is a device that produces a very intense beam of light. The beam can be so powerful that it can cut through skin, bone, and even metal. Lasers have only been around for a short time, but they have gradually crept into many areas of our lives. Everything from supermarkets and music systems, to holograms and weapons systems uses laser technology.

Early laser

In 1960, U.S. scientist Charles H. Townes (1915–) and a team of researchers from Columbia University in New York built a laser made from helium and neon. The laser looked a little like a neon light tube, but it had special mirrors at each end. As electricity passed through the gas-filled tube, electrons collided with the helium and neon atoms. Eventually, the helium and neon atoms started to glow—just like a normal neon light. The difference was that the laser emitted only red light. Trapped between the two mirrors, the laser light bounced back and forth, making the helium and neon atoms glow even more. Once the light had become powerful enough, it emerged as an intense beam of laser light. The light was even more powerful since the mirrors focused it into a very tight beam. This made it very easy to direct and control.

Uses of lasers

Helium-neon lasers have many different uses. They are found in supermarket barcode scanners, compact disc (CD) players, and laser light shows. In medicine, laser technology is now used to analyze blood cells. Scientists count blood cells by

DID YOU KNOW?

DIFFERENT LASERS

Helium-neon lasers are not the only lasers that use the noble gases. Argon and krypton lasers are also common. Everything from eye surgery and the manufacture of electronic chips for computers, to holography and laser light shows, use argon and krypton lasers. Excimer lasers use reactive gases such as chlorine and fluorine mixed with argon or xenon to produce ultraviolet laser light. Other lasers use helium mixed with the vapors of the metals mercury and selenium. Although these are not as popular as the other types, helium-mercury-selenium lasers are used to inspect computer chips, in printing, and also in chemical research.

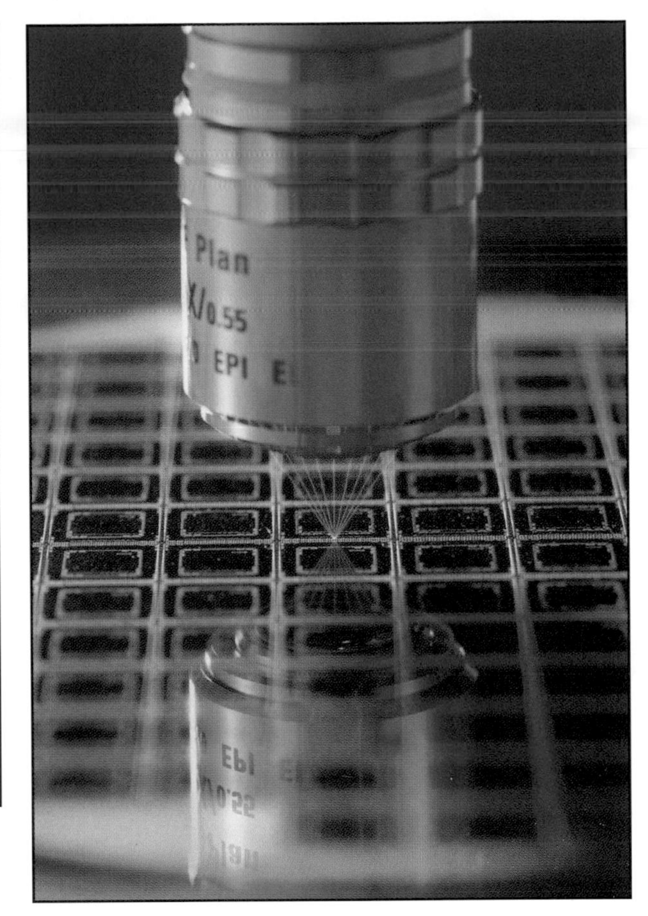

High-resolution laser beams pass over tiny electrical circuit boards to check for surface damage.

mixing the blood with a chemical that attaches special markers to the cells. When laser light is shone on the blood, it makes the markers glow. By measuring how much light is given off, scientists can work out how many blood cells there are. In industry, engineers use lasers to line up tools. They rely on the fact that a laser beam travels in a straight line. A laser is mounted on one part of the tool. A sensor is mounted on the other part of the tool to pick up the beam. Only when the two parts are exactly in line will the sensor detect the laser beam.

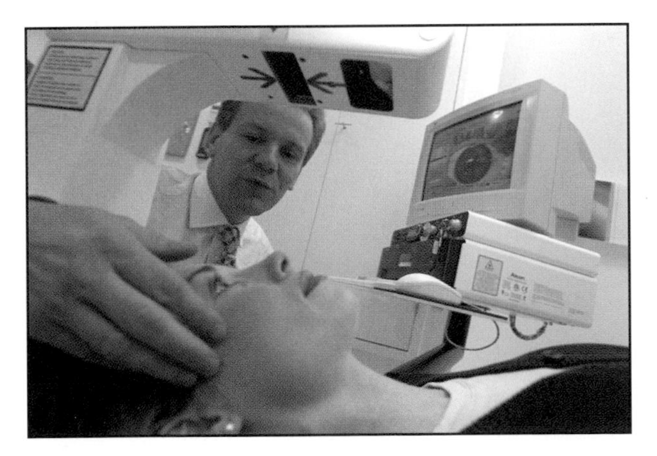

Laser eye surgery is a nonintrusive way of correcting problems with eyesight. The surgeon controls the power of the laser by altering the focus of the beam.

Unreactive atmospheres

The reluctance of the noble gases to react with other chemicals may seem like an undesirable trait. But there are many different situations where it is important to keep different chemicals from reacting, and nothing does this better than a noble gas.

A mixture of argon and krypton is used to make an unreactive atmosphere inside an incandescent lightbulb. As electricity flows through the tungsten filament, the metal gets very hot and glows. At the same time, tungsten atoms escape into the bulb. If the lightbulb was filled with air, then the tungsten atoms would react with oxygen in the air, and the filament would quickly

NOBLE GAS FACTS

There are many uses for noble gases as unreactive atmospheres:

- To remove carbon from steel to make stainless steel.
- To provide a clean and unreactive atmosphere for the manufacture of electronic chips for computer equipment.
- To help chemists carry out reactions that would be dangerous or impossible in air.

A metalworker uses a metal inert gas (MIG) welding torch. A noble gas such as argon is used to surround the molten metal and protect it from attack by oxygen in the air.

SEE FOR YOURSELF

You can test the effect of an unreactive atmosphere for yourself. Put a candle in a glass jar and light it. (Ask an adult to help.) Then put a lid on the jar and watch what happens. You will notice that the flame quickly dies away. The flame uses up all the oxygen in the air, leaving nitrogen, carbon dioxide, and noble gases in the jar. Since these gases are less reactive than oxygen, the candle cannot burn, and the flame dies out.

burn away. Lightbulbs filled with the mixture of argon and krypton last much longer. These gases do not react with the tungsten atoms and are not changed by the intense heat produced by the filament.

Welding

One very important use for argon is in welding. In this process, two pieces of metal are joined by melting them together. Welding is extremely dangerous. If the metals get too hot, they can react with oxygen in the air and start to burn. To stop this from happening, argon gas is used to "shield" the metals from oxygen in the atmosphere. As the metals are welded, argon is continuously pumped over the area. The argon pushes away the surrounding air, removing the dangerous oxygen atoms. Since argon is so unreactive, it does not react with the metals, which can now be welded together safely.

A worker at a nuclear plant in Idaho uses an argon- and helium-filled glove box to prepare radioactive sodium fuel rods.

Helium underwater

Did you know that the air you breathe can sometimes be poisonous? Most people never experience this unusual effect. But normal air can be extremely dangerous for divers who explore the depths of the ocean.

When people dive deep underwater, the water around them places huge pressure on their lungs. The pressure can be so strong that their lungs may collapse. To stop this from happening, divers breathe air at an equal pressure as the water around them. When air in the lungs is under pressure, however, more oxygen and nitrogen atoms are forced into the bloodstream. Excess nitrogen in the blood can make divers feel like they are drunk, while the extra oxygen causes them to shake uncontrollably. These are symptoms of an illness known as decompression sickness, or the "bends."

Divers who work at great depths usually breathe a mixture of helium and oxygen. Helium replaces nitrogen because it is light, easy to breath at high pressures, and does not harm the body. The mixture contains less oxygen than normal air, but the body still gets enough to survive.

An underwater diver breathes a mixture of helium and oxygen to dive to a depth of 200 ft (60 m).

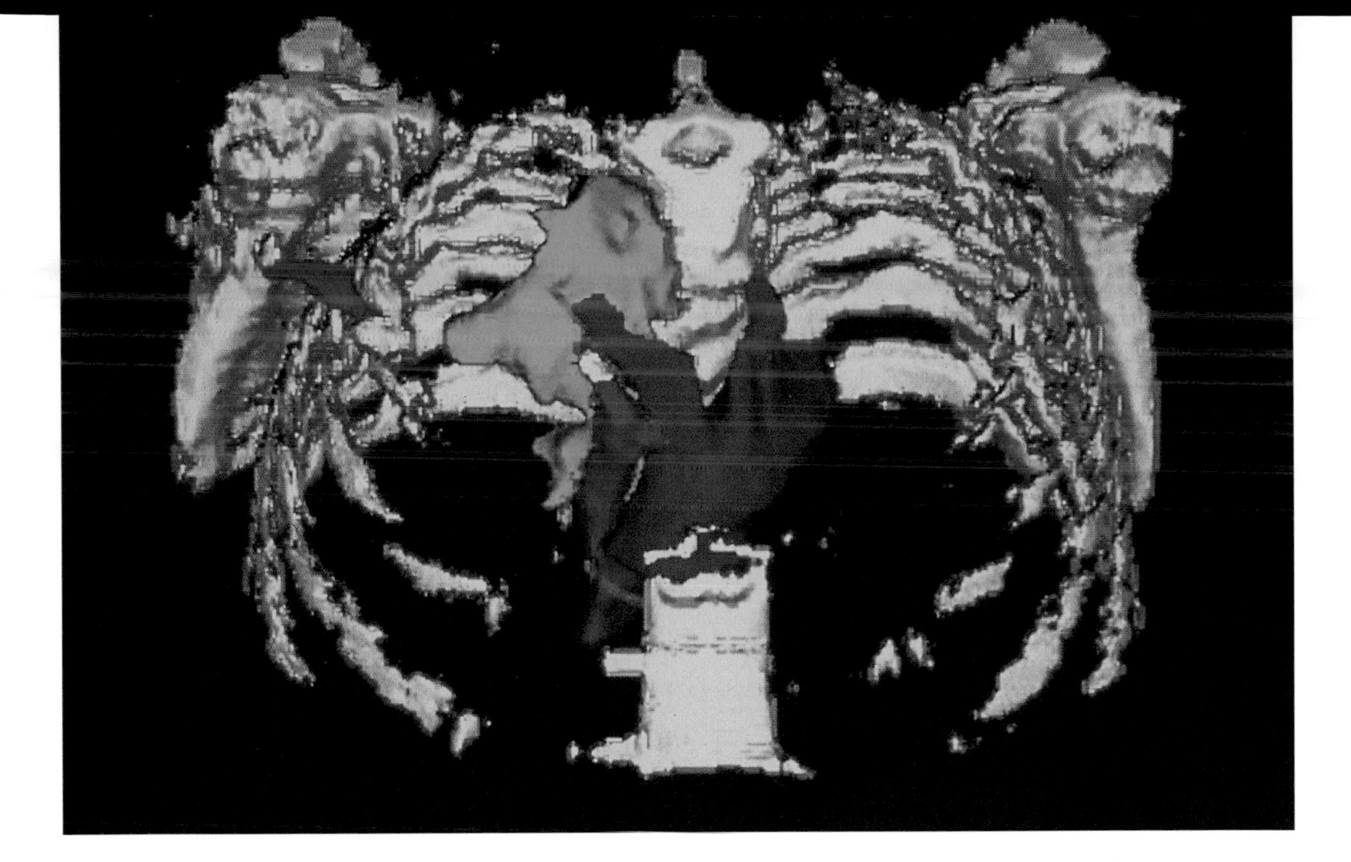

Radon in medicine

A false-color computed tomography scan shows a pancreatic tumor in green. The red and blue areas are the main blood vessels supplying the tumor. Tumors such as these respond well to radon radiotherapy.

Radon is naturally radioactive, and it is usually regarded as a serious health hazard. That is why it may seem strange that this element can be used as a treatment for cancer.

Cancer is characterized by the abnormally rapid division of living cells. These cells may form a lump called a tumor. Tumors and cancerous cells can spread and disrupt normal body functions. If cancer is left unchecked, it is often fatal.

Doctors have found that radioactivity can destroy the rapidly dividing cells of some types of cancer. If doctors place a small capsule of radon gas inside a tumor, the radioactive decay eventually destroys the tumor. Since radon is only radioactive for a few days, the capsule is harmless to the rest of the body.

DID YOU KNOW?

RADON AT HOME

Radon is a product of the radioactive decay of radium. People who live above underground deposits of radium are at risk of exposure to radon gas. This has been linked to lung cancer. Luckily, radon meters are used to measure the radon levels in domestic houses. It is usually easy to stop radon from seeping into homes.

Periodic table

Everything in the universe is made from combinations of substances called elements. Elements consist of tiny atoms that are too small to see. Atoms are the building blocks of matter.

The character of an atom depends on how many even tinier particles (called protons) there are in its center, or nucleus. An element's atomic number is the same as the number of its protons.

Scientists have found around 110 different elements. About 90 elements occur naturally on Earth. The rest have been made in laboratories.

All these elements are set out on a chart called the periodic table. This lists all the elements in order according to their atomic number.

The elements at the left of the table are metals. Those at the right are nonmetals. Between the metals and the nonmetals are the metalloids, which sometimes act like metals and sometimes like nonmetals.

- On the left of the table are the alkali metals. These elements have just one electron in their outer shells.
- On the right of the periodic table are the noble gases. These elements have full outer shells.
- Elements in the same group have the same number of electrons in their outer shells.
- Elements get more reactive as you go down a group.
- The number of electrons orbiting the nucleus increases down each group.
- The transition metals are in the middle of the table, between Groups II and III.

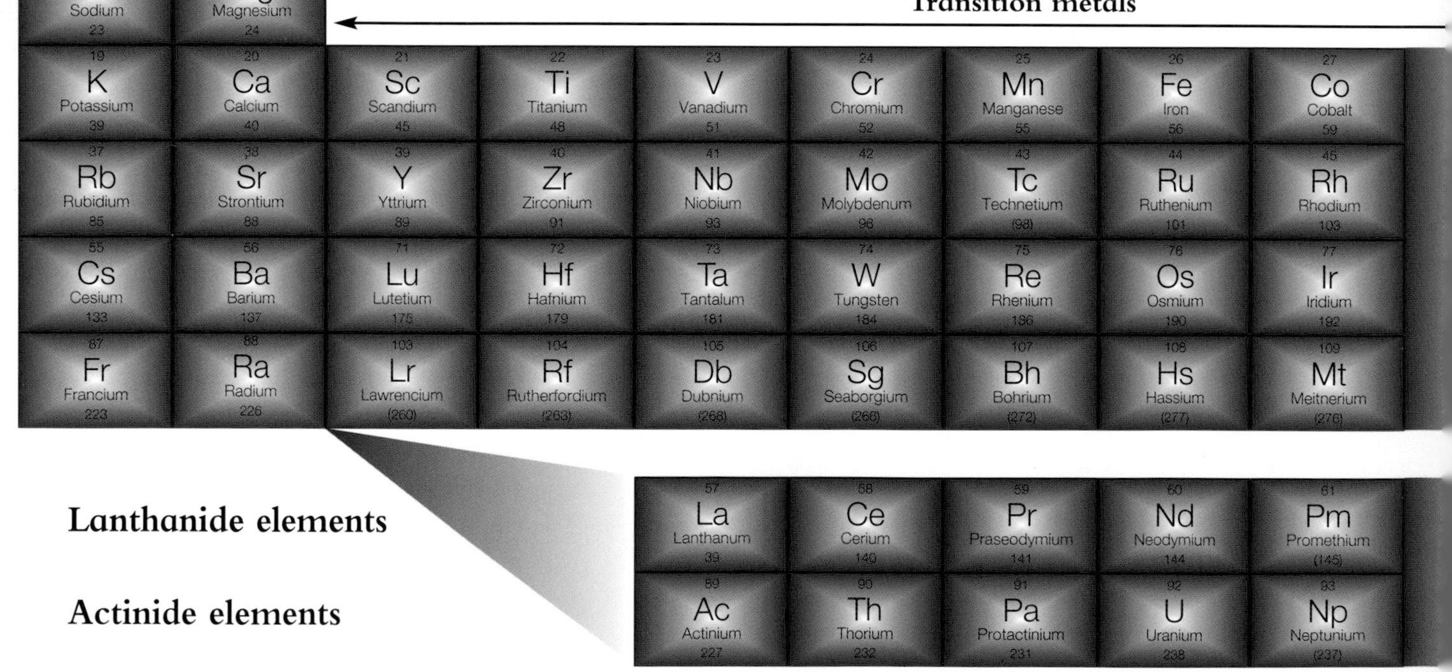

The horizontal rows are called periods. As you go across a period, the atomic number increases by one from each element to the next. The vertical columns are called groups. Elements get heavier as you go down a group. All the elements in a group have the same number of electrons in their outer shells. This means they react in similar ways.

The transition metals fall between Groups II and III. Their electron shells fill up in an unusual way. The lanthanide elements and the actinide elements are set apart from the main table to make it easier to read. All the lanthanide elements and the actinide elements are quite rare.

Noble gases in the table

The noble gases form Group VIII of the periodic table. Each Group VIII element has a full complement of electrons in its outer shell. Atoms form chemical bonds with other atoms by trying to fill up the outer shell. Since the noble gases have a full outer shell, they do not form bonds with many other atoms.

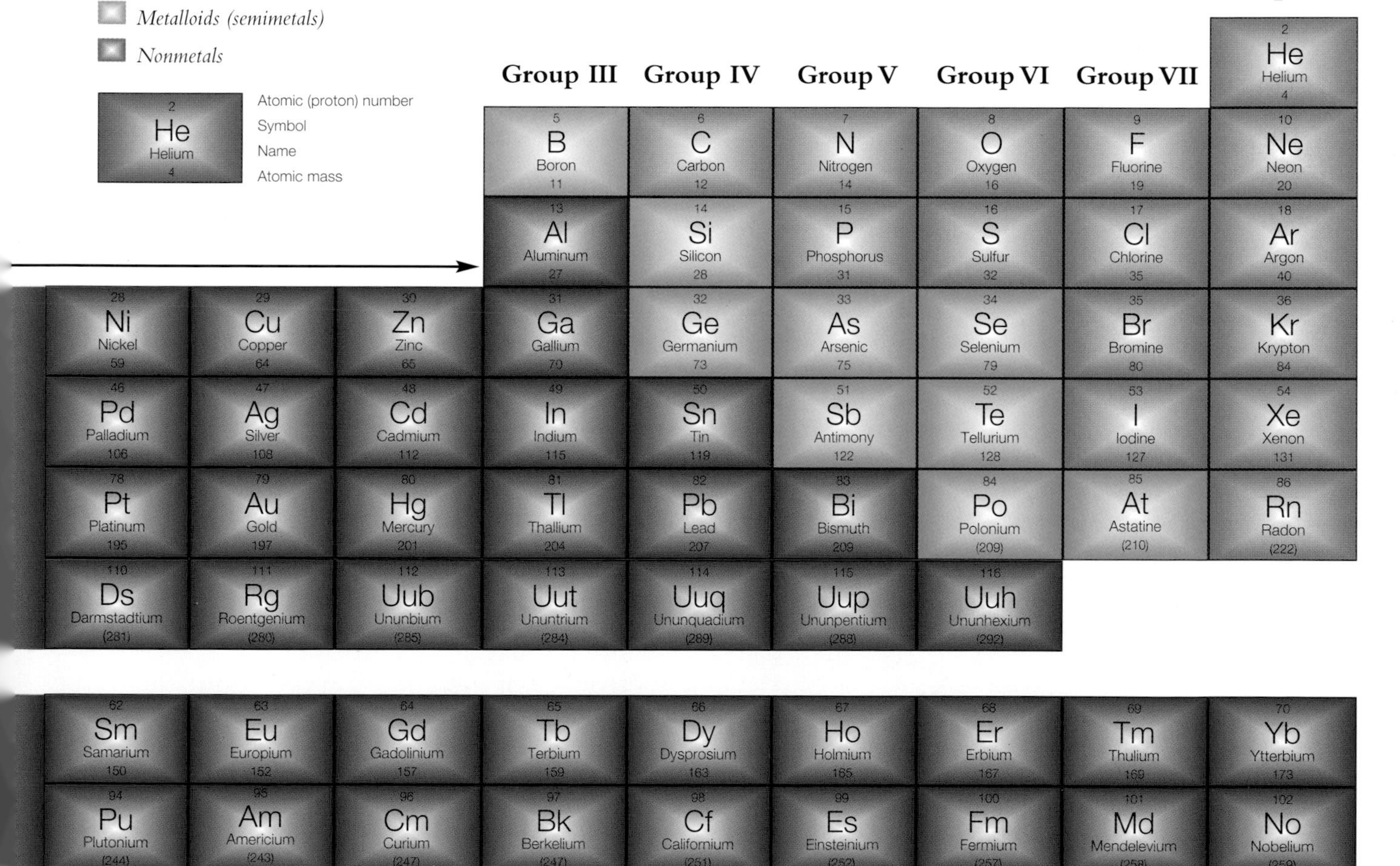

Chemical reactions

Chemical reactions are going on all the time—candles burn, nails rust, food is digested. Some reactions involve just two substances; others many more. But whenever a reaction takes place, at least one substance is changed.

In a chemical reaction, the atoms stay the same. But they join up in different combinations to form new molecules.

Crystals of krypton fluoride (KrF_2) form when krypton gas is passed over liquid fluorine at a temperature of −320.8° F (−196° C).

ATOMS AT WORK

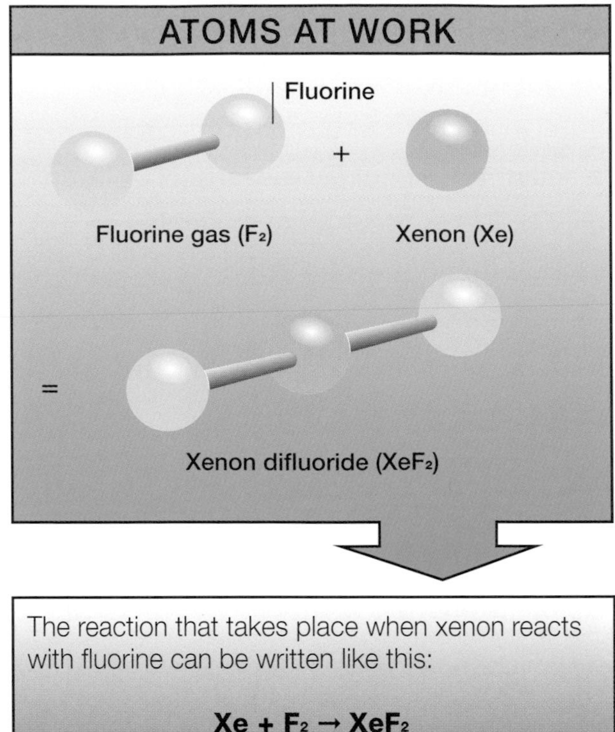

The reaction that takes place when xenon reacts with fluorine can be written like this:

$$Xe + F_2 \rightarrow XeF_2$$

Writing an equation

Chemical reactions can be described by writing down the atoms and molecules before the reaction and the atoms and molecules after the reaction. The number of atoms before will be the same as the number of atoms after. Chemists write the reaction as an equation. This shows what happens in the chemical reaction.

Making it balance

When the numbers of each atom on both sides of the equation are equal, the equation is balanced. If the numbers are not equal, something is wrong. A chemist adjusts the number of atoms involved until the equation does balance.

Glossary

absolute zero: The coldest possible temperature. Absolute zero is 0 Kelvins (–459.67° F, or –273.15° C).
atom: The smallest part of an element having all the properties of that element.
atomic mass: The number of protons and neutrons in an atom.
atomic number: The number of protons in an atom.
boiling point: The temperature at which a liquid turns into a vapor. The noble gases have very low boiling points.
bond: The attraction between two atoms that holds them together.
compound: A substance made of two or more elements that have combined together chemically.
corrosion: The eating away of a material by reaction with other chemicals, often oxygen and moisture in the air.
distillation: A process that separates a mixture of different liquids with different boiling points.
electrode: A material that is used to conduct an electrical current to or from an object.
electron: A tiny particle with a negative charge. Electrons move around the nucleus in layers called electron shells.
element: A substance that is made from only one type of atom.
ion: An atom that has lost or gained electrons. Ions have either a positive or a negative electrical charge.
isotopes: Atoms of an element with the same number of protons and electrons but different numbers of neutrons.
laser beam: A beam of high intensity light waves with the same wavelengths.
metal: An element on the left-hand side of the periodic table.
molecule: A particle that contains atoms held together by chemical bonds.
neutron: A tiny particle with no electrical charge. Neutrons are found in the nucleus of every atom except hydrogen.
nonmetal: An element on the right-hand side of the periodic table.
nucleus: The dense structure at the center of an atom.
periodic table: A chart of all the chemical elements laid out in order of their atomic number.
products: The substances formed in a chemical reaction.
proton: A tiny particle with a positive charge. Protons are found inside the nucleus of an atom.
radioactivity: The release of energy caused by particle changes in the nucleus.
reactants: The substances that react together in a chemical reaction.
ultraviolet: A form of radiation similar to light but invisible to the naked eye.
viscosity: A fluid's resistance to flow.

Index